I0479492

Parker's Solar Quest

In the depths of space, a spacecraft so bold,
A solar explorer, its tale will be told.

1st Edition

www.versatileread.com

Document Control

Book Name	:	Parker's Solar Quest
Document Edition	:	1st Edition
Document Release Date	:	14th May 2024
VR Product Code	:	20241201PSQ

Copyright © 2024 VERSAtile Reads.
Registered in England and Wales
www.versatileread.com

All rights reserved. No part of this book may be reproduced or transmitted in any form or by any means, electronic or mechanical, including photocopying, recording, or by any information storage and retrieval system, without the written permission from VERSAtile Reads, except for the inclusion of brief quotations in a review.

Feedback:
If you have any comments regarding the quality of this book or otherwise alter it to better suit your needs, you can contact us through email at info@versatileread.com
Please make sure to include the book's title and ISBN in your message.

About the Authors

Ayesha Saeed

Ayesha Saeed, currently pursuing a Bachelor's degree in Artificial Intelligence, stands as a cosmic champ, her fascination with the universe as infinite as the data points in a neural network. With a series of national Astrophysics victories and experience as the Astronomy Head for multiple science Olympiads, Ayesha's academic pursuits in AI complement her astral expeditions. She aspires to ignite the curiosity of future · celestial adventurers, weaving poems and pictures to inspire wonder. When she isn't developing code on her laptop, you can find her crafting whimsical tales that led to her debut book, 'Parker's Solar Quest'.

Laiba Mohsin

Laiba Mohsin, an undergraduate Computer Science student, has a mind wired for both digital wizardry and celestial exploration. After guiding the astronomy module to victory in the National Science Bowl and participating in several NASA competitions, she's on a mission to make space education as accessible as stargazing on a clear night. Her thoughts often wander into the abyss, contemplating the mysteries of black holes and the cosmic phenomena they entail. This passion drives her to support the cause of astronomy in her homeland of Pakistan, starting with 'Parker's Solar Quest'.

The combined expertise of both women, propelled their friendship to NASA contests and finally, co-authoring 'Parker's Solar Quest'. Research has been the focal point of their writing, guiding you through the star studded universe with the assured hand of true explorers. Their journey is sure to inspire a comradeship in the cosmos.

ON THE RACE TO THE SUN....

SUN'S DISTANCE FROM EARTH

149 million km

Ever wonder why we know so little about our favorite star, the Sun?

This is because the human eye can only observe a certain amount of things.

Looking at the Sun for too long can cause permanent eye damage.

This is why you should wear sunglasses!

WHY IS THE SUN SO IMPORTANT?

In the sky so high, the
Sun shines bright,
Gives us warmth and brings
us light.

From tiny bugs to
creatures grand,
The Sun's energy helps
life expand.

Sunlight brings nature
to life,
helping **ecosystems** thrive.

For without the Sun,
you see,
Life on Earth wouldn't be!

The Parker Solar Probe is a NASA spacecraft that helps us solve mysteries about our star which have intrigued us for many years!

One of these mysteries is why the Sun's outer layer, called the **corona**, is hotter than its surface, the **photosphere**.

It's like solving a big puzzle, and this probe will help us find the missing pieces.

SUN'S STRUCTURE

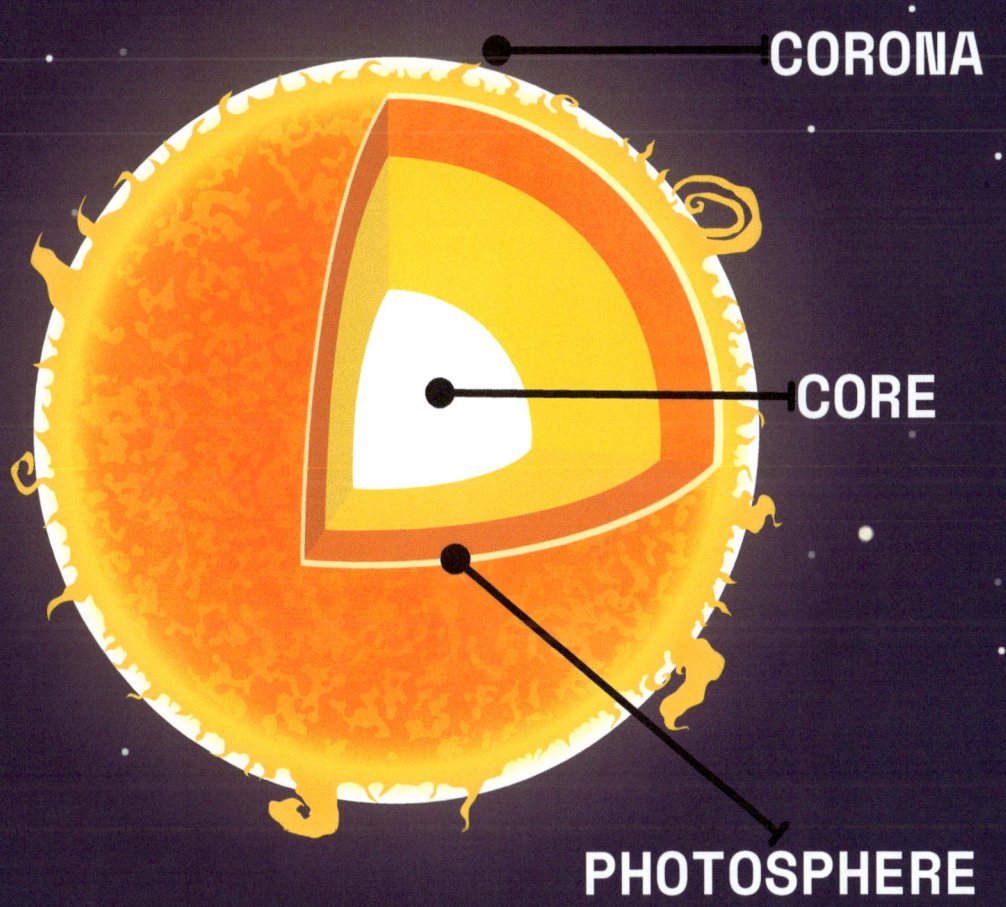

CORONA

CORE

PHOTOSPHERE

PARKER HELPS SCIENTISTS:

1) Follow the path of energy that makes the Sun's corona hot and creates a fast stream of particles called the '**solar wind**'.

2) Study the stuff called 'plasma'.

This teaches us about the shape of the **magnetic fields** around the Sun and how they change over time.

3) Explore how particles near the Sun get really fast and travel around in space.

THE LAUNCH

A powerful rocket launched the Parker Solar Probe from Cape Canaveral in Florida on August 12th, 2018.

This began Parker's incredible journey to study the Sun up close.

DID YOU KNOW?

The Parker Solar Probe is the fastest man-made object, traveling at 0.05% the speed of light!

WHY IS IT CALLED PARKER?

Parker Solar Probe, its
name so bright,
Let's learn why it shines
with such might.

Named after a scientist,
wise and keen,
Dr. Eugene Parker, helped
us explore the unseen.

With his discoveries, he
paved the way,
To understand the Sun's
powerful display.

With each discovery, it
brings back,
Dr. Parker's legacy on
its track!

TOUCHING THE SUN

To help the probe get closer to the Sun, it uses a clever trick involving Venus.

Parker takes advantage of Venus' strong pull to slow down and change its path.

This move helps the probe get into a special **orbit** that brings it closer to the Sun for its mission.

In its 8th **flyby**,
Parker entered the
solar atmosphere.

This is when it
reached its
closest point to
the Sun called
the '**Perihelion**',
only about 6.16
million kms away
from the actual
surface!

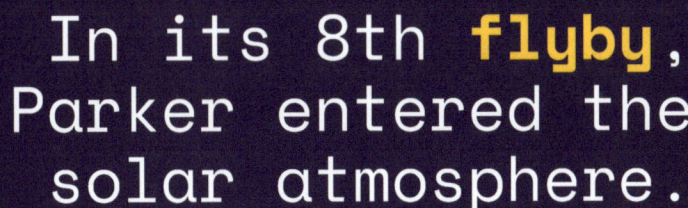

This helped Parker break
records in speed, and
distance from the Sun!

PARKER'S DISCOVERIES

Through journeys close to
the Sun's embrace,
Parker discovered marvels
in every space.

Magnetic fields performing
a celestial play,
Revealing their power in a
marvelous way.

Solar winds, so profound,
Particles racing, with
energy unbound.

Parker witnessed their
patterns, their flow,
Unraveling the secrets
they're eager to show.

PARKER'S DISCOVERIES

Coronal loops, like
cosmic flames,
Twisting and turning, with
mysterious aims.

Microflares erupting, in
bursts of pure delight,
Unveiling the Sun's glory,
shining so bright.

Parker's keen eye caught
each dazzling flare,
Illuminating the Sun's
radiant lair.

In the depths of space, a
probe did roam,
Uncovering the Sun's
secrets, making
them known.

DID YOU KNOW?

Parker discovered a long tail of particles escaping from Venus. It stretched almost 8,000 kms!
This could tell us how Venus lost its water, becoming dry and uninhabitable.

WHY DOESN'T IT MELT?

If the scorching Sun can melt your ice cream from millions of kilometers away, how can Parker touch it? Why doesn't it melt?

To start off, we need to learn the basic differences between Temperature and Heat.

TEMPERATURE VS HEAT

Temperature is the speed of moving particles

whereas...

Heat is the energy that transfers between particles

At 20°C, you will feel hotter in:

a packed classroom full of students

more heat transfer

rather than...

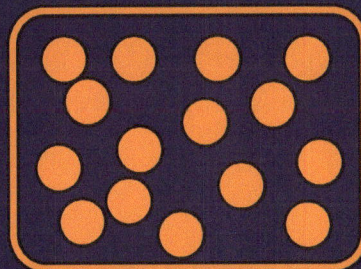

a playground far away from your friends

less heat transfer

You can keep your hand in a hot oven longer than a pot of boiling water.

This is because the water particles are more closely packed.

(Do not try this at home)

Since the Corona is less dense (fewer particles) than the **photosphere**; Parker bumps into a few hot particles, experiencing high temperatures and less heat.

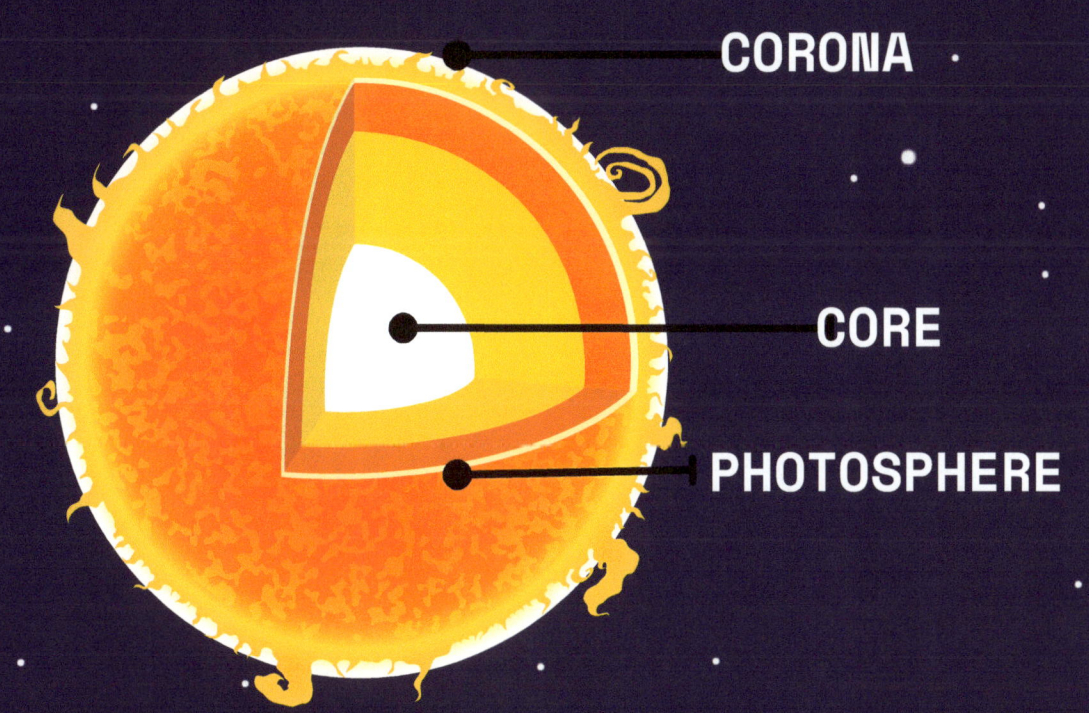

CORONA

CORE

PHOTOSPHERE

DID YOU KNOW?

Over 7 years, the spacecraft will complete 24 orbits around the Sun!

The Parker Solar Probe has a special feature; the **heat shield**. This is made up of layers of material that can resist heat.

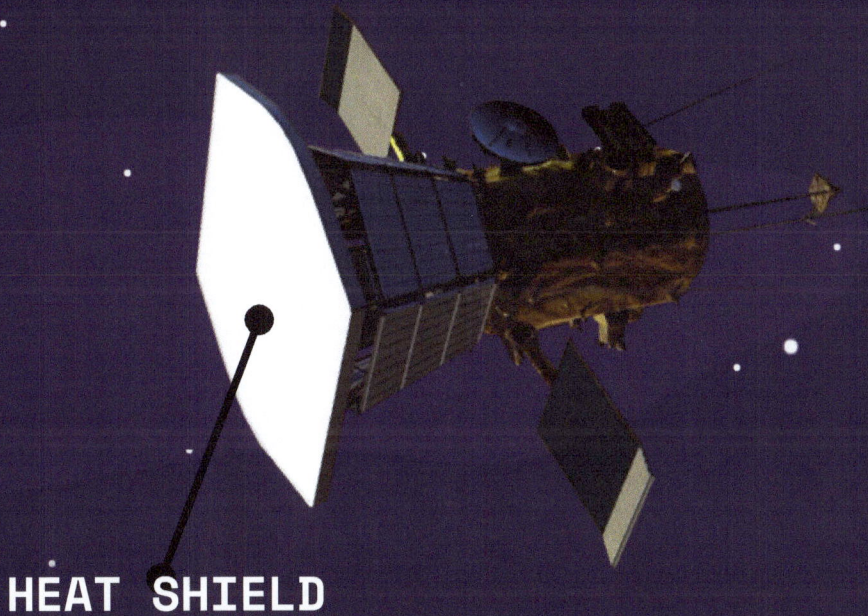

HEAT SHIELD

It also has a carbon layer sandwiched between 2 **carbon-carbons**.

This carbon is made up of something called **graphite epoxy** (material found in golf clubs and tennis rackets) which is lightweight and can be heated up to 1400°C.

Graphite Epoxy

The fascinating detail is that the **carbon-carbon** at the back remains cool even at high temperatures!

One of the layers of the heat shield is a white **ceramic** coating. This is the outermost layer that faces the Sun.

White helps reflect light and absorbs less heat.

This is why you would prefer to wear white on a hot summer day!

Carbon-Carbon
Thermal Shield

Cooling System

Heat Shield

Solar Arrays

PARKER'S HELPING HANDS

.As we move away from its
fiery glow,
Why does the Sun's heat
continue to grow?

Faraday's Cup, our guide
through space,
Reveals why the Sun's heat
finds no trace.

In the space where solar
winds blow strong,
A metal cup traps **plasma**,
where it belongs.

Catching **charged particles**
in its way,
Faraday's Cup ensures
they stay.

Solar Arrays stretch
out wide,
Capturing sunlight as
their guide.

Motors move with
silent grace,
In the shade, they find
their place.

Leading edges face the
Sun's light,
Fueling the spacecraft's
flight.

While the **solar limb sensors**
catch the Sun's rays,
The **heat shield** adjusts to
find the right ways

Autonomous and free, Parker
charts its own course,
Moving ahead with
unwavering force.

As we journey through space,
our knowledge expands,
Thanks to Parker and its
helping hands.

DID YOU KNOW?

When a space probe gets really close to the Sun, its shield faces temperatures as high as 1,400°C, but its instruments stay cool at around 30°C!

So far, Parker has helped us find new doors to space exploration.

Now you can be the key to these doors and unlock more mysteries of the universe.

It was curious little minds like yours that led to the Parker Solar Probe's spectacular journey!

So kids, look up to the
stars shining bright,

Imagine a spacecraft, a
brave, daring sight.

And learn from the
probe, courageous
and true,

To reach for the
stars, there's no
limit for you!

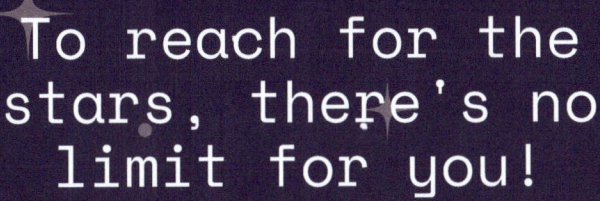

...THE PARKER SOLAR PROBE
WON!

GLOSSARY

Autonomous: being independent and able to do things on your own. (31)

Carbon-carbon: a super strong material made from layers of carbon fibers stuck together with a special glue. (26,27,29)

Ceramic: a hard, brittle material made by heating clay or other non-metallic minerals. It is often used for its ability to withstand high temperatures. (28)

Charged particles: tiny bits of matter that have electrical energy. (30)

Core: the central region of a celestial body, like a star or planet. (7,23)

GLOSSARY

Corona: the outer layer of the Sun's atmosphere. (7,9,24)

Coronal Loops: loops of gas on the Sun's surface, shaped by its magnetic field. They are visible during solar activity and can stretch far into space. (18)

Ecosystem: a group of living organisms that live and interact with each other in an environment. (6)

Flyby: when a spacecraft passes close to another celestial body without entering its orbit. (16)

GLOSSARY

Graphite Epoxy: a strong, lightweight material made by mixing graphite and epoxy glue. (27)

Heat Shield: a special protective layer made of carbon-carbon. It shield's the spacecraft from the Sun's intense heat. (26,28,29,31)

Magnetic fields: invisible areas of magnetic force. (10, 17)

Microflares: tiny bursts of energy and heat on the Sun's surface. They are like mini versions of larger solar flares, releasing smaller amounts of energy. (18)

GLOSSARY

Orbit: the curved path an object takes around another object in space due to gravity. (15,25)

Perihelion: the point in the orbit of a planet or a comet when it is closest to the Sun. (16)

Photosphere: the surface of the Sun, a term which means "sphere of light". It is where the light we see comes from. (7,8,24)

Plasma: the fourth state of matter, after solid, liquid, and gas. Think of it like a soup of charged particles. Lightning and the Sun are examples of naturally occurring plasmas. (10,30)

GLOSSARY

Solar Limb Sensors: Long finger-like devices on the edges of the probe to help it maintain proper orientation and keep it safe from the Sun's light. (31)

Solar Winds: streams of charged particles flowing from the Sun into space. (9,17,29)

Uninhabitable: an environment where it is impossible for living things to survive. (18)

www.ingramcontent.com/pod-product-compliance
Lightning Source LLC
Chambersburg PA
CBHW042028230526
45474CB00006B/42